Endorsements for "The Last Trumpet" and Charlotte Bergmann

In *The Last Trumpet*, Charlotte Bergmann sounds the alarm and extends an urgent invitation for America and the world to return to the principles that the Lord sets forth. Please read the book and answer the call.

— **Evangelist Alveda King** *Executive Director for Civil Rights for the Unborn and niece to Dr. Martin Luther King, Jr.*

Charlotte Bergmann is a champion of the Judeo-Christian ethic. In her new book, *The Last Trumpet*, Charlotte captures the pain of the past and molds it into a bridge of healing and conversation promoting the salvation of America. *The Last Trumpet* is an important read for every American in these uncertain and tumultuous times in our nation.

— **C. L. Bryant** *Pastor, FreedomWorks Senior Fellow, Nationally Syndicated Radio Host of The C. L. Bryant Show, Filmmaker (Runaway Slave) and Author (A Race For Freedom)*

It's a blessing to be associated with God-fearing patriots like Charlotte Bergmann, and I'm grateful for her doing her part to promote truth! She's not writing from theories, but personal experience. She's endured hardship but didn't *allow* the policies of pity, seduce her into being a taxpayer-funded career victim. She leans on the Lord and enjoys His blessings of independence. Her book, *The Last Trumpet*, is a helpful light to liberty while exposing the party of dependency.

> — **Alfonzo "Zo" Rachel,** *Social and Political Commentator, known for his unique and profound perspective on national issues, Founder of Bronze Serpent Media who has shared the stage with the likes of Andrew Breitbart, Herman Cain, Larry Elder, and Bill Whittle among many others.*

The Last Trumpet

In the Battle for America

Charlotte Bergmann

with Pastor Charlana Kelly

The Last Trumpet
In the Battle for America

Copyright © 2019 by Charlotte Bergmann
with Pastor Charlana Kelly.

Published by SpeakTruth Media Group LLC, Crockett, TX
www.speaktruthmedia.com

All rights reserved. No part of this publication may be reproduced, stored in a retrieval system, or transmitted in any form or by any means—electronic, mechanical, digital, photocopy, recording, or any other—except for brief quotations in printed reviews, without the prior permission of the author and publisher.

Book cover design by Leslie Kinney for SpeakTruth Media Group LLC in collaboration with Charlotte Bergmann.

Front & back cover author photos credit: Patricia Possel
Front cover illustration photo credits: ©Richard Hausdorf/123RF.COM & ©Lukas Gojda /123RF.COM.

ISBN-13: 978-0-9985190-7-4 (pbk.)

Printed in the United States of America
First Edition: June 2019

Vs. Steve Cohen

Dedication

To the memory of my grandson, Brandon.

During the writing of this book, my grandson was murdered in Memphis. He was shot in the back, after refusing to get into an altercation with another man.

Brandon's light was a beacon of hope. He had dreams and desires. His future was bright. It seems ironic that he fell victim to the very things I am fighting with all my heart, mind, soul, and strength to end. He was a peacemaker, a unifier, a lover of God, life, and people.

While Brandon's light no longer burns, mine shines brighter with a higher purpose and determination to fight in his memory for the people of this nation to experience the blessing of our foundational truths that I hold dear—life, liberty and the pursuit of happiness.

Table of Contents

Foreword ix

Introduction xv

A Voice for the Voiceless 26

The Incurable Wound 35

The Diabolical Plan 44

They Think We're Stupid 55

A House Divided 66

Honoring the Truth 72

The Last Trumpet 82

Dedicated & Determined 94

End Notes xcix

Foreword

It was 2014, the Producer of my Podcast, *A Voice for Our Time*, suggested guests for me to consider. Charlotte Bergmann's name was on the list. When I learned she was running for a seat in the US Congress; I immediately knew I had to speak with her.

On June 25, 2014, Charlotte and I began a journey of Divine intersection as the Lord stirred my heart to pray for her and gave us opportunities for brief encounters in and around Memphis.

I was excited to interview Charlotte that day and interested in her political platform. So, I wanted to hear what she had to say about the issues plaguing America. It became abundantly clear during the interview that Charlotte was a woman who possessed the courage of her convictions, and the determination to speak up for those who could not speak for themselves. She had a passion to do her part to open the eyes of those who were deceived and living beneath their potential. She clearly had the wisdom necessary to lead and bring unity and peace to both the people she would represent and our nation.

At the time of our interview, the nation was divided; today, it is polarized. And the gap is widening! Now more than ever, we need to build a bridge together that will both unify people and save our nation.

I believe we would all acknowledge that in addition to the division, our nation has experienced severe moral decay over the past few decades. We need healing and restoration, and it will take more than a Band-Aid® to bind up our gaping wounds. But, if we are going to heal, we must go after the root causes of our decline. And, quite frankly, I believe the root is in the church.

We might all agree it is sin, but sin is only the symptom of a once powerful church that slowly and consistently compromised the truth.

While I don't want to place blame on the church as an institution, I will boldly assert the real issue is the voices that represent the church. Voices from high places in the church that now routinely demand, elevate, and celebrate sin. Voices that have largely remained silent over the past decades, shying away from the arenas of influence that can shift this nation back to God, and ultimate peace and prosperity.

The compromise has been breathtaking. The silence has been deafening, literally advancing Hell's plan. And, those who are silent are complicit as well.

Thank God we do have bold Christian leaders who continue to speak out, albeit they are far and few between. But, imagine with me for a moment the power that would be released if church leaders and believers all took a stand for righteousness at the same time. Power so consuming that unbelieving Americans would fall to their knees in fear. He desires unity, Jesus prayed for unity, and by faith I believe, we will see it happen in God's way and time.

We need godly leaders in every arena. What a blessing it is to have them in government, an institution established by God himself. And Christians, you need not forsake your responsibility to impact the very institution that God ordained.

Charlotte Bergmann is not only a Christian impacting the arena of government; she is also a politician. She is God's daughter, who loves people, and wants the best for families. She is not afraid to give her voice to speak the truth that needs to be heard. She is a herald of our time delivering a Divine message to all who have ears to hear. I hope you glean wisdom from her as you read this book and allow her words to sink deep into your heart and change your mind if need be.

As we got acquainted during that interview in 2014, I noticed her heart and message were different than a lot of politicians I've had the honor of getting to know. She was more concerned about the restoration of God's ways, and protecting the foundation of Judeo-Christian values from where our nation's roots spring. I ended our interview in an unusual manner, but now I believe it to be significant. As I committed to pray for Charlotte, I said, "I feel like I've known you forever." And indeed I believe, it was preordained that we would meet and partner together in the plans of God for our nation.

Over the years since, her passion has not waned but intensified. She is fearless! Especially in the face of evil, and determined to represent both God and the good of the people.

We need more leaders like Charlotte Bergmann. Her message and voice bring clarity and peace.

In this book, together, we laid out the truth in hopes all who read its words will wake up and become the person God created them to be—a voice for what is right and true in the eyes of God alone.

Charlotte has fought valiantly since 2010 to be a voice for the voiceless, and I have no doubt she will continue until victory has been granted and the people are free as God intended them to be.

So, I say to you as you read this book, let the TRUTH be known and the rejoicing begin. The truth will set you free and that's exactly what Charlotte Bergmann is delivering to you here in *The Last Trumpet in the Battle for America*.

Pastor Charlana Kelly
Author, Speaker, TV Host, and CEO of SpeakTruth Media Group LLC

Introduction

I hear a trumpet sounding my friend, and I believe it could be the last blast. It's time to turn around.

Few times in history have so defined a nation's culture, and world events as the events of 1968. With breathtaking blows, each moment set a course for America's future, from the culmination of The Civil Rights Movement to the assassinations of Dr. Martin Luther King Jr. and Robert Kennedy. There were sharp contrasts in speech, like the Black Panther's call for violence

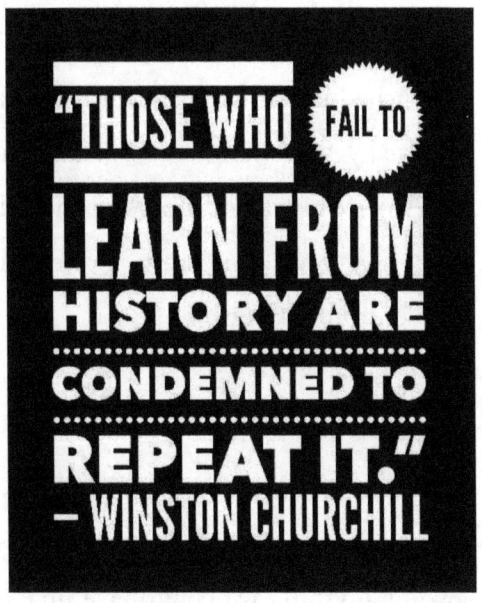
"THOSE WHO FAIL TO LEARN FROM HISTORY ARE CONDEMNED TO REPEAT IT." – WINSTON CHURCHILL

and the peaceful pursuit of equality through Dr. King's emboldened speeches. Their words were either like fuel to the fire of those demanding retribution, or healing balm that moved the heart of a nation towards peaceful change.

It was a time of challenge to every structure in America, both governmental and societal. Promiscuity, drugs, and the rock and roll craze sent youth into a downward spiral of unconventional and destructive lifestyles. Where sex with whomever, whenever, wherever was celebrated as the sexual revolution opened the door to disrupt traditional family values. An all-out rebellion against parental norms and established laws for decency was unleashed to such a degree that many felt emboldened to demand a radical change to the American way of life. And then there was the Vietnam War that was violently protested at the Democrat Convention in Chicago.

The nation was in flames.

Riots, racial unrest, deadly attacks on law enforcement, and political upheaval all set the stage to shift our culture beyond anything we could have imagined. Anarchy wanted to rule, but peace and unity won, albeit a short-term win for what has proven to be a long-term battle.

Today we are witnessing the same undercurrent of 1968, proving that the hearts of men have not healed, as the same demands rise again through attempts to divide people by race and socio-economic status, defiance of and deadly force against law enforcement, student rebellion, and attacks against every structure. As this mindset of destruction has resurfaced, it is stronger and more destructive. People who adhere to this type of thinking are not only calling for a radical shift in culture but are also demanding a governmental shift to Socialism.

What is Socialism? At first glance, it is portrayed to be the answer of all society's woes, but as implementation unfolds, it is little more than a complete take-over of a free people who become wards of the state. Government control of every facet of life is the end-game. What people don't initially realize is that the ultimate intent of Socialist leaders is to remove God from society. At this point, every political, cultural, and familial structure, as we know it, is destroyed.

I was 14 years old at the time of Dr. Martin Luther King Jr.'s assassination. It was a pivotal time in Memphis, Tennessee as I would learn over the years. For me, though, I lived in a quiet, family-oriented neighborhood. I attended school with my friends, and we enjoyed the same things most children of that era enjoyed, even playing outside till late into the evening. I lived in a home with both parents and had a life free from violence. As I look back, I recognize we lived a very sheltered life hidden from all the turmoil going on in the nation. We were extremely poor, but I didn't know it. We were blessed to grow up with a strong foundation of God's Word, that centered around the church. My father was a Baptist preacher, who required we all memorize Scripture. I'm so grateful for those years. I remember the moral fiber of families was solidly Christian, and it was scandalous for a girl to get pregnant out of wedlock. Times have changed, and not for the better.

On March 22 of that year, I remember a strange snowfall that came to Memphis. The flakes were so big and beautiful, accumulating on the cars, rooves, and lawns, but quickly melting as it hit the street. I learned later that this was the day Dr. King was supposed to be in Memphis to offer his voice for the sanitation workers who were on strike, but he postponed his visit due to weather. He returned on the 28th but was hindered again, when the rally he was to attend turned violent. His third attempt proved deadly. I remember that day vividly because our neighborhood sirens sounded, indicating an emergency curfew when everyone was required to remain indoors. I lived on Chelsea Street at that time, which was a busy thoroughfare. I peeked out to see what was going on, the street was empty, and an eerie darkness filled the air. I will never forget it.

While I graduated from high school before the implementation of busing in my neighborhood, I do understand the effects of uprooting children from a peaceful habitation and forcing them into an unfamiliar place where in some cases they were not welcomed. Think of it for a moment. Think also of change to their daily schedule based on the time required for busing and family hardships that prevented parents from being actively involved in their child's education. Now children became aware of their poverty; they got to see what it looked like to have better things. Envy and greed set in, violence began to unfold, as frustration and confusion took root. At this point, I'm in high school, and I remember for the first time seeing black and white students fight one another.

When we demand rights that God has not given to us, we open a door for Satan to come in and wreak havoc in the lives of otherwise happy human beings. The enemy knows this well; the key to the defeat of any structure is instability. Chaos creates *instability* by either circumventing or removing the rule of law. Law brings order and order brings peace. When neither law nor order exists, we live in anarchy; meaning an absence of government or law — total mayhem results. The only hope is that people choose to govern themselves, which is a founding principle of our nation. America was and is a "Great Experiment" to determine if free people can govern themselves based on a set of moral laws without the need for government intervention. Samuel Adams once said, "While the people are virtuous, they cannot be subdued, but when once they *lose their virtue* then they will be ready to surrender their liberties to the first external or internal invader." Today, every moral code is being challenged, removed, or eradicated. Many have lost their virtue.

It must be stated that freedom does not come from a governmental structure. Every individual has an innate desire to be free. It is a gift given by a gracious Creator, and no one can remove that sense of liberty from the human heart. It cannot be legislated or enforced; it can only be protected; guarded and ensured by what is right and just, good and noble. Because there is a *higher authority* than the government, "faith" is under attack today. God simply cannot exist in the hearts and minds of people who want their government to be their provider.

In America today we are witnessing a magnification of good and evil. We are also seeing a dichotomy unfold, where people are calling evil good and good evil. Nowhere is it more evident than the issue of abortion. One group stands for the rights of the baby in the womb, while the other group stands for the rights of the woman. The core issue is life, the value of human life. The destruction of innocent life is nothing new; civilizations millennia past have sacrificed their babies and children seeking a blessing from false gods. The "false gods" we sacrifice our children to today are selfish ambition and greed, or the love of money.

When one group is fighting for life, freedom, security, and peace, it's obvious their values and pursuit line up with a moral fiber that is congruent with God's desire for humanity. The other group is fighting for the right to murder, the destruction of foundational truth, and the devaluing of self and life. And, it's a proven fact that these matters are destructive to women, men, families, and the future of our nation. I take solace in knowing that the truth will always win. It will demand a hearing, and eventually, hearts are softened and changed as truth is realized. But, I can't help but wonder, have we gone too far as a culture?

The demand for rights is good and well, however like anything else those demands that start out to be virtuous can end up in depravity. Much like the recent laws passed to terminate the life of a child after a live birth; a living, breathing baby waiting to be "humanely" murdered at the request of its mother. Killing a baby after birth *is* infanticide! Ask Pennsylvania abortionist Kermit Gosnell who was charged with eight counts of murder; one for a woman who died after an abortion procedure and seven babies[1] partially born alive who were murdered by snipping the spinal cord. He is serving life in prison. Should his sentence now be commuted because a nation's leaders have decided acts like his were just?

I am often reminded; love does not demand its way or rights unless there is injustice, and then the demand is only too right the wrong. Dr. Martin Luther King Jr. was doing precisely this—trying to right the wrongs, in a spirit of unity, love, and peace. And, indeed, an innocent baby is worth demanding rights and justice. Who will speak for them? Who will stand on their behalf?

There are many gray lines today, many more than in Dr. Martin Luther King, Jr.'s day. Where there's a gray line, we must choose to be on the right side of that gray line. And we must realize that refusing to take a side on the issue is taking the wrong side. Silence is complicit; it's as though we performed the act ourselves. John Wesley said, "What one generation tolerates, the next will embrace." We've tolerated so much over the last 50 years, looked the other way, judged the people involved, then slowly embraced it and tolerated it so much that we now celebrate it. The consequences of our cooperation will be too high, my friend. I promise you; it will cost us more than we are willing to pay.

After the events of 1968, I witnessed families break apart, the escalation of violence, and the disintegration of the moral fiber of our nation. Children began to live in fear of violence, and pregnancy among teens and single women began to rise. The dependence of generations, once strong and vibrant, upon a menial government stipend, held them back and kept them in bondage as a new plan unfolded to both marginalize and eliminate them.

In that tumultuous year, chaos overtook our nation in what seemed to be a point of no return. However, during times like those America always rallies as most Americans reject the ideas and social engineering that attempt to take the nation further and further away from the foundational truths that structure our way of life.

We all want good for our children and grandchildren. So why do we support anything or anyone who is destroying a future of success for them? Why do we look evil in the face and call it good, and vice versa? Could it be we have been swayed by creative deception, a bomb wrapped in beautiful detail, a wolf in sheep's clothing? Is it because the *Left* tells us what we want to hear? Could it be the *Left* is falsely accusing *conservatives* of doing what the *Left* has been doing for decades? Could it be they have done exactly what they set out to do, deceive you? Manipulate you into believing they are right? After all, isn't that how to destroy the thing that stands in the way of gaining control? Accuse the innocent of being guilty to turn the hearts of those who need to be deceived? Once the loyalty of the delusional is gained, eliminate the innocent so they cannot uncover the truth.

We seem to find ourselves again at a similar moment in history. But the stakes are higher, the deceived have increased multiple times over. They now demand the very things that will destroy them. And as a result, our nation is teetering on the edge of a cliff about to go over.

I hope that the message in this book will dispel the myths, debunk the lies, and release a fresh perspective of truth to set you free and give you permission to think for yourself. And, receive the wisdom you need to make choices that will bring freedom and prosperity to you and your family.

It is evident, the decisions we've made up to now have not produced the outcome we want for our children. In this book, I will release truths we must all accept if we are going to save ourselves and especially our children and grandchildren from the destruction looming if we don't turn around.

Like any other issue that clouds our thinking, we must be courageous enough to step out of the fog and look for ourselves. Silence the voices that stir strife and division, then search for the facts. When we do, we will find the truth. It's liberating to be your own man or woman, to know what you believe and why you believe it. And, be able to back it up with facts. Don't let your voice be part of a choir of dissent, let your voice rise above the crowd to sing singularly with the One who created the universe and has good plans for His people that when released will bring peace and hope.

CHAPTER 1

A Voice for the Voiceless

Truth: *Dr. Martin Luther King Jr. was not the man revisionists and opportunists portray him to be.*

Throughout history, voices have risen to release a message that brings deliverance, whether to a person, a group, or a nation. Abraham Lincoln rose to bring an end to slavery. Nelson Mandela rose to end Apartheid. Dr. Martin Luther King Jr. rose to end segregation.

I've often wondered what America would look like today if Dr. King were able to live out his days. I can't help but believe it would look very different. He was a man of peace who wanted to bring a nonviolent end to inequality. He promoted unity and loved all people regardless of their race.

Today his heart, spirit, and message have been hijacked by people who use his legacy to further their agenda. Bold statements are made like, "He was in favor of Socialism and universal income." Or, "He would support abortion today." Individuals in power who want to associate themselves with his legacy, believe that by making such assertions they can manipulate the heart of the hearer as they take Dr. King's words out of context to make their point.

While history often repeats itself, certain aspects of history cannot be compared without understanding the culture, political climate, and societal norms of that time. To compare the times that Dr. King lived in, to today, and thereby the context from which he spoke is both misguided and misleading. In the 1950s & 60s, the treatment of black Americans was extremely harsh. Severe inequalities existed, and segregation was a real issue that needed to end. Neither exists today; however, they are often invoked as a reminder to inflame the hearts of people and hold them in bondage to the past.

When Dr. King spoke of "socialism," he wasn't speaking of a government structure that rationed food, destroyed business, and flattened income opportunity. He believed in "the American Creed," a good work ethic and the ability of his people to rise to equal status with any affluent person regardless of the color of their skin. Dr. King spoke of an ideology based on Christian values, charitable deeds, peace, and unity, where people of all races would have equal opportunity to achieve the American Dream. He was not and would not today be in favor of people being paid to do nothing and add nothing to better society.

When he championed economic rights for black Americans, he was not advocating universal income or dependence on social programs. He was pointing out the disparity between job opportunity and income potential. His dream was "deeply rooted in the American Dream." One of prosperity and freedom, not subservient dependence upon a system that intends to keep impoverished people poor.

And as to abortion, Dr. King's niece, Dr. Alveda King, shared during an interview at Notre Dame Law School's pro-life student organization in 2012, "Dr. King would never have agreed with the violent violation of the civil rights of the millions of aborted babies. And Planned Parenthood's subsequent blitz of women's health problems related to chemical and artificial birth control methods."

While Dr. King received the Margaret Sanger Award in 1966, it was not widely known at that time the truth about the woman who pioneered a method to control the population of whole groups of people, who she deemed societal "weeds" as she mentioned numerous times in her writings. Abortion was not legal at the time of his award, so Planned Parenthood's efforts were focused on other forms of birth control. To say that Dr. King would support abortion today is disingenuous and deceitful to further a cause no doubt Dr. King would find abhorrent if he were alive today.

These insinuations are lies released to deceive and manipulate the minds of people. Can't you see it? If you can, are you willing to overlook it? Dr. King would not. He would stand up and speak the truth, especially for those least among us.

Regarding his political party affiliation, some say that he was only a registered Republican because that was the only option for black voters at that time. What they don't acknowledge in their rhetoric is the fact that the Democratic Party would not allow black Americans to register as Democrats. More telling if you think about it! Why? Because it was the Democratic Party that wanted to suppress their voting rights, they sought to silence, segregate, and extinguish the black race altogether.

Many say Dr. Martin Luther King, Jr. was a Civil Rights Leader and Activist, but I believe he was a prophet. He was not a politician and did not serve the desires of political opportunists or schemers. He served God and honored God's ways as he went about speaking and sharing the truth with passion and perseverance until the end.

One of his closest confidants Minister Abernathy once said, "You may be assured that we won't ever let your words die. Like the words of our Master, Jesus Christ, they will live in our minds and our hearts and in the souls of black men and white men, brown men and yellow men as long as time shall last." Not only did those closest who understood the heart and humanity of Dr. King not allow his words to die, neither did his God who he served.

His words carry weight today for those who dream of an America where there is an opportunity for all. A nation of unlimited opportunity is the America we have today because of Dr. King's sacrifice. And those who can hear the truth, recognize the truth. If they are brave enough, they will realize how much Dr. King's words and courageous actions changed the future of our nation.

Since I lived during the time leading up to Dr. King's assassination, this is what I know. We cannot compare his day to ours, although without his efforts and sacrifice segregation would still exist. I have lived in the freedom he fought for, and I am not willing to give it up to bondage.

Life was not perfect in 1968, but it was more centered around the family and parents were very much involved in the lives of their children raising them in the fear of the Lord with strong Christian values as their foundation. Today we live in a culture far worse because of the continued intentional destruction of the moral fiber and the rule of law in our nation.

Over the decades, we have witnessed the deterioration of the family, absence of fathers, drugs, abortion, escalating violence, entitlement attitudes, and dependence on the government. The *Left* has been successful in their efforts to keep the same people Dr. King fought to free, enslaved to an ideology of reliance. The *Left's* ideologies are the antithesis of the truth he fought and died for; every human being is created equal. All have inalienable rights given by their Creator through birth alone, and all have the same God-ordained gifts to use to provide for themselves, their family, and their community.

It is foolishness to apply Dr. King's words to today's political debate other than to say he would want liberty, unity, equality, and compassion for all.

Dr. Martin Luther King, Jr. took a stand for what was right in the sight of God. He made his choice to be a voice of peace and encouraged others to do the same. He wanted what God had given to all, and because of his stand, this nation changed. Great strides were made toward unity and togetherness.

The voices of dissent and division today are not working for God's will. They are fulfilling the agenda of Hell. The only way those voices can be successful is if we allow their hatred and lies to penetrate our hearts and minds like poison.

The liars will get their due portion. Do not be afraid of them and do not be silent. Be bold and fearless for truth. Be courageous in seeking it out and let the Holy Spirit guide you. After all, He is the Spirit of Truth given to teach, lead, and guide us.

But above all, let's not be counted among those who intentionally deceive to control. Like Dr. King, let's take our stand with God and further the agenda of Heaven which is liberty through Christ, peace, and love for all, and a determined singleness of mind and purpose as we lead others to Him.

We need the voice of Dr. King today, but that voice will have to come from someone else. Will that voice, be you?

Dr. King wanted more for you; he had the Father's heart. He had a dream; he dreamed that you would prosper. He dreamed that your future would be full of promise and peace. Will you dream with him and become an advocate for what is good and right and pure and noble?

If you join with his dream, then deliverance, healing, and restoration must come. And, you must align yourself and your heart with the One who gives life abundantly.

It's time to let go of the past and step into the future with faith and hope that everything promised to you by your Maker is available for you and those who you lead. Let's let go together and lay hold of the things that make this nation great.

CHAPTER 2

The Incurable Wound

Truth: *Your enemy refuses to let the wound heal.*

To separate yourself from the past so that you can move into the future, you must allow your wounds to heal.

Most wounds heal quickly, but the wounds of slavery and racism in America seem incurable. Why?

There's a straightforward answer. Someone keeps picking at the wound. Whether that's you or someone else, the wound is prevented from healing by the constant poking and disruption of the natural process to heal completely. Here's the thing, from your enemy's perspective, this wound is not meant to heal. Please read that again. The wound *is not intended to heal*. It has become a weapon. And the weapon is wielded intentionally upon you every single time "someone" wants to move you in a predetermined direction to hold you in bondage.

Doesn't that make you mad? Doesn't that make you feel like a puppet in the hand of an evil master?

While racism has existed throughout the history of man, America abolished slavery on February 1, 1865, as President Lincoln approved a Joint Resolution of Congress, which was vehemently opposed by the *Democrats*. The truth is that racism will always exist. Every race and creed deal with this type of prejudice in some form or fashion. Christians themselves are discriminated against to this day. In many nations, they are considered minority second-class citizens only useful for menial service. Today multiple thousands are enslaved and imprisoned around the world.

In America today, most people have no problem with those of different races and socio-economic backgrounds. Across this land, people of all kinds live in harmony with respect and love for one another. There are curious pockets of people in specific locations that continually bring up the race issue. And what's the response when racism is uttered? Anger, violence, demands for retribution! Friend, it's a false narrative that must be kept in focus for the *Democrats* to maintain control.

Who are they controlling and why? The purpose of their control is power. The *Democrat Party* must get and maintain the support of the people they need to control so that they can remain in a place of power. They manipulate the minds of people using old means and slanderous language to incite rage and violence. This cycle keeps the wound from healing when there is no reason for the wound to be open. Aren't you tired of fighting a paper dragon? What do I mean by that? There's nothing there. It's smoke and mirrors: lies and false accusation to incite.

Who is the accuser of the brethren? Satan. Why do we attack and devour each other to the bone, then proclaim the name of Jesus and shout hallelujah every Sunday? Aren't you tired?

Now about that horrible word, racism, you need to know this. It is used as a weapon to silence the truth-teller or those who dare get out-of-line by challenging the status-quo as truth is revealed. Because we have been unknowingly deceived, we become the lynch men. Like Saul of Tarsus, persecuting the church and martyring its members as he unwittingly tried to destroy God's kingdom. Satan is the author, and we need to stop dancing with him.

Here's the rub; the *Left* can't allow you to believe that you are equal, respected, and valuable to the community. They would rather deal in delusion to keep you suspicious and angry. Keep you begging them for crumbs. They can't allow unity, because to unify would mean you see that those who supposedly abhor you actually love you. You would see that prosperity is for you too.

Unity releases blessing from Heaven. Disunity releases destruction from Hell. It seems to me that we ought to be doing everything within our power to be at peace with one another and live in unity together. Still, we choose to segregate ourselves. Realize this; we CHOOSE to separate and look at everyone else with an eye of distrust. We choose the wrong way. And, the fruit of it has come upon us.

There is an old saying about falsely accusing your enemy. It comes from a playbook written by a personal mentor of some on the *Left*. "Ridicule is a man's most potent weapon."[2] Indefensible ridicule silences the opponent so that there is no opportunity for an honest discussion. It isolates and divides. It puts the accused on the defensive and creates instability that may never return to normal.

The mentor's name is Sal Alinsky, who blatantly said he would "organize hell" and dedicated his book, *Rules for Radicals*, to Lucifer who he said was the greatest *organizer* of all time. Why would the church follow this man? Why would a Christian pastor or minister follow his playbook? I don't for a minute, believe it is intentional. I do think a Christian who follows this strategy doesn't know who this organizer *really* is.

The same is said of Hitler. Christians followed him unwittingly; Stalin too. How is it we can be deceived to the degree that we not only believe the lie, but let it have power over our hearts, minds, and relationships? How is it we can be so blind? The truth was right in front of us, yet we choose to believe the lie and take the wrong side.

The truth is that the *Democratic Leadership* are your real enemy. And until you realize the people you think are there to help are destroying you, your family and your future, you will follow those *leaders* to your death. It's your enemy who will keep your wounds gaping wide open. Your friend will help you heal, find peace, contentment, health, and prosperity. Your friend lifts you, while your enemy keeps you victimized in a constant state of numbed confusion and dependence. You don't help a poor man by keeping him poor.

And we are so confused and on edge emotionally that whenever there's the slight bit of thought an event has happened with racial undertones, we immediately cry racism and end up rioting, and destroying our neighborhoods. Only to find out a short time later that the event was not what we thought it was at all. The opposite was true. This cycle happens over and over again. When will we stop the madness and start reasoning with one another about the truth?

Most recently, Jussie Smollett who staged a supposed racial and homophobic attack on himself in Chicago by two men who said, "this is MAGA country,"[3] wrote threatening letters to himself, then put a noose around his neck allowing two friends to rough him up a bit before he called to file a false report with the police. His inflamed accusations thought to be a hate crime were proven to be little more than a selfish, conniving man who wanted attention.

This event was a potential powder-keg that could have unleashed an explosion of riots not only in Chicago but throughout the nation. This false accusation[4] had the potential to incite violence over race, the LGBT movement over the homophobic slant, and almost 63 million people who voted for President Trump in 2016 with the MAGA reference.

The results, if this false report had gone unchecked by the facts, could have been more devastating than the riots of 1968. Many people could have been killed, and indeed, the division between groups in our nation would have widened significantly. We may not have recovered from it.

Why are we so quick to believe every charge of racism? Why can't we be rational and wait for all the facts to come out? There's always a rush to judgment, and lives are destroyed as a result.

Thank God the facts were quickly uncovered in the Smollett case, saving Chicago and this nation from further mayhem.

The fact is racism is not a skin issue; it's a sin issue. The heart of man cannot be regulated or legislated. Only God can intervene. So, we have to do things God's way if we want the blessings He promised. Wrong responses will never bring an end to racism. Allowing the accusations and opinions of talking heads to whip us into a frenzy will not solve anything; it only serves to perpetuate the wound.

But, when you understand the deviant nature of why the *Left* wants to stir strife and contention, you have to decide to stop responding the way they want you to. You must become a part of the solution, rather than helping to inflame the problem.

Those of us who dare to be God's voice in this hour need to boldly stand up and confront the lies with truth. We need to have the wisdom and discernment to see the spirit behind the manipulation. We need to tell others the truth and help them find freedom. We need to lead with a focus of bringing peace to every situation and teach others to be slow to anger.

Racism will never cease. So, our response to it must be Christ-like. As Christians, it is imperative that we be unifiers and peacemakers. When we are mistreated, and we believe it is because of our race, we should offer a hand of mercy and kindness to those who are not yet walking in the liberty of Jesus. It's up to us to change our mindset and see like Him rather than seeing everything the way our enemy wants us to view it and ourselves.

As we begin to extend mercy and kindness, our nation will heal. As we look beyond the faults of others and walk in peace with all people, unity will be realized. As we do these things, our children will have a chance at a successful future. We've got a lot of work to do. A lot of mindsets need to change. The best place to begin is with our own heart and mind. We are in this together, let's make the choice today to become who God created us to be in this great nation we have the blessing of calling our homeland.

These deceivers I mention have been at work for almost 100 years. It's time to recognize them and stop being ensnared by their plans.

CHAPTER 3

The Diabolical Plan

Truth: *We have become victims of genocide by our own hands. — Reverend Johnny Hunter*

Margaret Sanger was a master of deception and delusion who used black ministers and medical professionals to pedal her plan to control the population of people she and her eugenics friends deemed "weeds" on society. If the black leaders resisted, she persuaded them with her pleas for the need to have family planning and healthcare for poor women to trick them into

Anatomy Lesson for Feminists

Woman's Body

Not a Woman's Body

A Baby is NOT a Female Body Part!
This future citizen of the world deserves rights and is not your own female organ to pick & choose.
Stop saying; "My Body, My Right!" when it's NOT your body!
A baby is an entirely separate person!
A precious human being! Not an elective organ!

agreeing she had the best interests of the black community at heart.

As she spoke to other groups of people, those of the eugenics persuasion, her words would detail her plan to extinguish the "negro" population. Much of which was overlooked for decades. Who is Margaret Sanger? She is the visionary founder of Planned Parenthood, a devilish woman, with a cold heart not only toward the least among us but toward the church and charitable organizations that helped the poor as well.

In her early writings, she blamed the benevolence of those who wanted to ease the suffering of others, saying, "Organized charity itself is the *symptom of a malignant social disease.* Those vast, complex, interrelated organizations aiming to control and to diminish the spread of misery and destitution and all the menacing evils that spring out of this sinisterly fertile soil, are the surest sign that our civilization has bred, is breeding and *perpetuating constantly increasing numbers of defectives, delinquents and dependents.*"[5]

Margaret Sanger is the women whose organization and successors are both celebrated and protected to continue the work of extinguishing the black race. The devastation has been significant, and it's time to dispel the myth that Planned Parenthood has good intent toward black and poor women today.

First, I want to define eugenics, so you clearly understand it. According to the Merriam-Webster Dictionary, "eugenics" is *the practice or advocacy of controlled selective breeding of human populations (as by sterilization) to improve the population's genetic composition.* A method developed to improve the human race but largely rejected in the 1940s because its doctrines were horrifically utilized by Adolf Hitler to justify extinguishing the Jews.

While Sanger increasingly used deceptive speech to trick her intended targets, her plan continues to this day with new names and the same cry for family planning and women's healthcare. It's like dressing up a pig. No matter how many costume changes the pig enjoys, it's still a pig. And, Sanger's diabolical plan against black women is still unfolding today, but in more horrific ways than we could have ever imagined when the "birth control" issue was debated throughout the first half of last century.

As she began her quest, the government was called upon for intervention. In 1926 Sanger said in a speech, "It now remains for the U.S. government to set a sensible example to the world by offering a bonus or yearly pension to all obviously unfit parents who allow themselves to be *sterilized* by harmless and scientific means. In this way, the moron and the diseased would have no posterity to inherit their unhappy condition. The number of the feeble-minded would decrease, and a heavy burden would be lifted from the shoulders of the fit."[6]

Her first "birth control" clinic was established in New York City in a predominantly black community known as Harlem. Over many years, more black babies have been aborted in New York City than are born alive. Doesn't that alarm you?

By 1939 her brazen efforts were boldly outlined in a plan to reach out to black leaders — specifically ministers — to help dispel community suspicions about the family planning clinics she was opening in the South.[7] She wrote, "We do not want word to go out that we want to *exterminate the Negro population*, and the minister is the man who can straighten out that idea if it ever occurs to any of their more rebellious members."

Reverend Johnny Hunter, the National Director of Life, Education, and Resource Network (LEARN) once said, "We have become victims of genocide by our own hands." If someone told you that the very thing you are fighting for the right to do is destroying your future, would you continue to fight for it?

Margaret Sanger is still having her way as Planned Parenthood steers clear of the fact, they are the largest provider of abortions in our nation preferring to point out that their services are primarily mammograms and pap smears, you know, women's healthcare. It's the same old lie Sanger realized would work to deceive minds and hearts as she set her course to eliminate those she deemed "undesirable." And, black leaders throughout this nation champion their cause, many among them Christians.

Recently Ohio State Representative Janine Boyd, a black woman, tried to exempt black women and their unborn babies from "heartbeat bill" legislation requiring abortions be performed before a gestational heartbeat is detected at about six weeks. She proposed in her exemption that black women could continue to have abortions up to 20 weeks. She went on to say, "Black slaves were once treated like cattle and put out to stud in order to create generations of more slaves," she said. "Our country is not far enough beyond our history to legislate as if it is."[8] She is talking about her own community as though they are ignorant slaves being forced to have children they chose to procreate. How misguided can one person be? Her Amendment failed in the vote. Thank God.

You must open your heart and mind to see clearly. Otherwise, you will fight for your own demise. Why would anyone support, champion, and celebrate a person whose sole intent for social change was to eliminate a race of people from existence? No man, no woman in their right mind, even a Christian would be a party to such an evil plan.

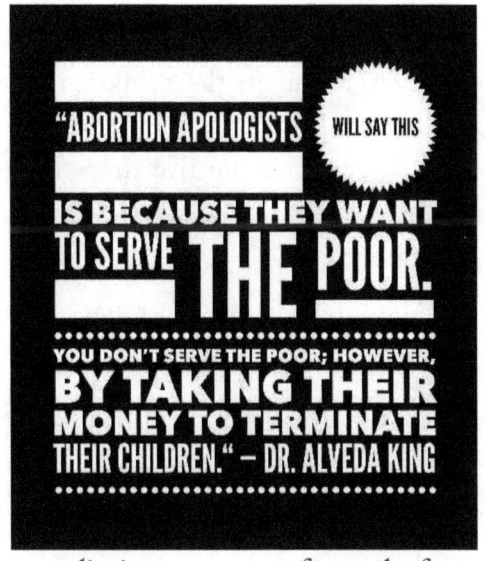

> "ABORTION APOLOGISTS WILL SAY THIS IS BECAUSE THEY WANT TO SERVE THE POOR. YOU DON'T SERVE THE POOR; HOWEVER, BY TAKING THEIR MONEY TO TERMINATE THEIR CHILDREN." – DR. ALVEDA KING

Many deny to this day that Planned Parenthood *targets* minority communities filled with low-income people. However, the statistics speak for themselves. A 2010 Census proved 79% of Planned Parenthood's surgical abortion facilities are within walking distance of minority neighborhoods filled with black and Hispanic women.[9]

Dr. Alveda King once pointed out, "A majority of abortion clinics are in areas with high minority populations. Abortion apologists will say this is because they want to serve the poor. You don't serve the poor; however, by taking their money to terminate their children."[10]

Black women are five times more likely to abort their children than white women.[11]

The CDC reports on a national level nearly half of all pregnancies among black women end in abortion. To contrast, only 11% of pregnancies among white women end with abortion. Could this be because black females have been targeted and deceived?

We are the ones who are choosing to allow our race to become extinct. And we are celebrating the very woman who said she would do it through "birth control." The Rev. Walter Hoye, founder of the Issues for Life Foundation, has spoken out about the consequences of high abortion rates among black women. He believes that if the replacement rate of births does not turn around, black Americans might become extinct within a few decades.[12] Tragic!

Let me paint a sobering picture for you. Most recent stats show more than 900 black babies are aborted each day in America. Hoye also notes, "Between 1882 and 1968, 3,446 blacks were lynched by the Ku Klux Klan (an arm of the Democratic Party of that day). Today, abortion kills more black Americans in less than four days than the Klan killed in 86 years!" [12] And, the intentional destruction of the black family has been the result.

The dismantling of the traditional family structure began in the 1960s during the sexual revolution as the right to use birth control became a woman's right to privacy under the law. Sex outside of marriage became the norm, swiftly changing how people viewed pregnancies that resulted from their sexual escapades and how children would be raised if they were *allowed* to live. As a result, fathers were no longer necessary in the family unit and for the first-time women assumed the role of single parent ready or not.

Women were also encouraged to have children out of wedlock, even incentivized to do so. Some say that the social policies of our government punish marriage, ensuring that the black woman becomes dependent on the government through subsidies for provision and care. By 1973, if a woman decided to terminate her pregnancy—the next level of birth control—she could do so legally. It wasn't enough to control the population of poor people through prevention methods, now we have empowered women to choose to kill their babies if they so desire. We are empowering them to murder.

God gave humanity boundaries. He shared with us what is good and right. He told us how to prosper and gave us promises accompanied by blessing. If we cross a boundary, we only set ourselves up to cross the next boundary, then the next, and the next. Every boundary we cross takes us further and further away from God. If we don't stop and turn around, we end up in a place where we can no longer hear Him or see Him. Sin in the heart of man disconnects him from God.

Condoms were available in the mid-1800s, and other physical devices used to suppress or prevent conception became available in the 1900s. In the 1960s birth control was revolutionized with the *convenience* of oral contraception. The 1970s delivered legalized abortion, which the *Left* originally termed, "safe, legal and rare." While most abortions take place in the first trimester, the time-limit keeps getting pushed further and further into the last trimester of pregnancy. Another boundary crossed. Now we are talking about aborting a child after live birth. Civilized society calls this barbaric. It is infanticide.

Are you going to look the other way?

Do not be fooled; God sees and knows and takes account. In fact, one of the things He hates, even detests is the shedding of innocent blood. Our soil is soaked with the innocent blood of babies aborted since its legalization; more than 60 million to date.

While one could say the sacrifice of children has gone on throughout history, it is happening right now on your watch. It is your time. Are you going to be silent while an entire race is intentionally extinguished?

The result of the demand for "birth control" is a complete and utter devaluing of human life that has translated far beyond the innocent baby in a mother's womb. If we don't value and protect the least among us, we will not value the family, our parents, the elderly, or any other human being. We won't value our nation, the freedom we enjoy, or those who fight to defend it. Think about this for a moment. It began by *first* devaluing of the poor.

Solomon wrote much about the treatment of the poor in the Book of Proverbs. It is obvious God's heart is toward the poor and those who *close their eyes to them receive many curses.* So, we must consider this, those who devalue life, devalue God, and scoff at those who believe. Just like Margaret Sanger said that she thought the church and charitable organizations who helped the poor were the roots of the problem. Are you beginning to see a correlation here? These ideologies all wrap back around Socialism and the mindset of control, extermination, and creating a race of people who are as Sanger put it once, were "thoroughbreds."

The poor, the church, and babies are all among Sanger's and the *Left*'s undesirables. If we don't wake up now and stand up for what is right in the sight of God, we are doomed.

Friend, they are banking on us remaining silent. We can't let them win! It's time to speak.

CHAPTER 4

They Think We're Stupid

Truth: *We're not—let's prove them wrong.*

Let's declare in unison, "I want to be poor, I want to depend on the Government, I want to be in debt to the third generation of my family, I hate my country, I hate my family, I want chaos, I want to walk wherever I go, I want to eat nuts and figs, I hate God."

Of course, we want everything the *Left* tries to convince us we need. When we resist, they throw out lines about "chains" and "reparations," they bait us, they switch us. Once they are secure in their power, they continue with their efforts to deny, obstruct, and destroy any freedom and liberty we might enjoy, then convince us we are better off when we are not.

Like when the last Administration said, "You can keep your healthcare plan, you can keep your doctor." Telling you what you want to hear knowing full well it is a lie. It was a lie. And, their lies and plan nearly destroyed the healthcare system in our nation. And, they are still not done, now they want Medicare for all! Who will pay for it?

Margaret Thatcher knew about Socialism. She said, "The problem with Socialism is that you eventually run out of other people's money."

We are watching the effects of Socialism unfold right now in Venezuela. No food, no medicine, 90% poverty rate, national debt the country cannot repay, and a leader who refuses to allow humanitarian aid in to help his citizens. The same is true of Cuba.

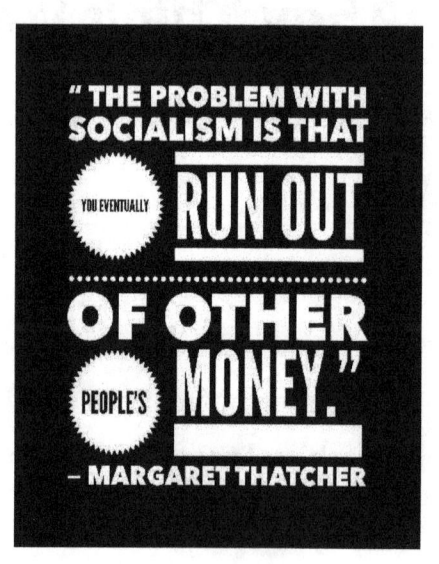

Socialists speak of their ideology as some sort of Utopian Society where every citizen has an equal share. Some even dare to say Jesus was a socialist. Not true! The only thing that is true about their "equal share" statement is that the citizens all have exactly WHAT THE GOVERNMENT WILL ALLOW. Government becomes the sole provider of all things; food, education, healthcare, reproduction, transportation, etc. And the scariest thing of all; government becomes "god."

It needs to be said here that if the government becomes *provider*, it means the government will *ration* to you what it thinks you need, nothing more but often much less. You lose all ability and right to decide for yourself. You lose your freedom right down to your basic necessities. Is that really what you desire?

The *socialists* don't want you to pull back the curtain on their plans. When you question them, they become defensive and attack your character rather than answering the question directly. Their foolish proposals all have one thing in common. Their plans will bankrupt the country and destroy our nation; 70% tax rates, no more cows, planes, trains, or automobiles. It's like an Odyssey. Are they kidding? No, they are not!

I must admit when this chapter title came to me, I was a little tentative with including it, but I think it needs to be said, "They think we're stupid."

Socialists promise you the world; like free this and free that. Tossing crumbs your way to keep your allegiance. Let me tell you something, free always costs you more than you are willing to pay; it could cost you everything, even your life.

They make deals with our enemies, sneak loads of cash out in the cover of night hoping no one will notice. Remember what the Good Book says, it seems right, sounds good, but the end of it is death.

In 2012, half the *Democrat Delegation* clearly booed including God and Israel in their Party's National Platform.[13] The Party Platform that year as written did not contain the typical language written to include references to God and God-given rights or affirming the role of Jerusalem as the capital of Israel as had been done in the past. Former Governor Ted Strickland proposed an Amendment to add this language where three times half the Delegation was obviously against adding both to the Platform. They say they are Christian, represent themselves as Christian when it is convenient and expedient for them, but their actions prove otherwise.

They show up in our congregations talking like they are from the "hood." Promising the world, delivering nothing. What has changed under their watch? What have they led that has ever turned into something good for you and your family?

Let's take a stroll down History Lane for a moment. Here's the inconvenient truth about the *Democrat Party* many are following.[14]

> *Since its founding in 1829, the Democratic Party has fought against every major civil rights initiative and has a long history of discrimination.* [14]

> *The Democratic Party defended slavery, started the Civil War, opposed Reconstruction, founded the Ku Klux Klan, [authored and enforced the Jim Crow Laws], imposed segregation, perpetrated lynchings, and fought against the civil rights acts of the 1950s and 1960s.* [14]

> *In contrast, the Republican Party was founded in 1854 as an anti-slavery party. Its mission was to stop the spread of slavery into the new western territories with the aim of*

abolishing it entirely. This effort, however, was dealt a major blow by the Supreme Court. In the 1857 case Dred Scott v. Sandford, the court ruled that slaves aren't citizens; they're property. The seven justices who voted in favor of slavery? All Democrats. The two justices who dissented? Both Republicans (Excerpt from Prager University, The Inconvenient Truth About the Democratic Party). [14]

You need to *know* this! Republicans DID NOT switch over to the Democrat Party. Believing that lie is beyond credulity. It is also an impossible feat. Why would *Democrats* support policies they fought a war to defend? That lie is at the core of why the wound of racism cannot be healed in America.

President Lyndon B. Johnson, a Democrat, despicably said, "I'll have them n*****s voting *Democrat* for two hundred years." [15] How? By deceit, by promising the moon, by false accusations and lying about the opposing party. And, I'd say he's been successful so far although the tide has been turning ever so slightly the last few years. And this is only one example of LBJ's contempt for black people, yet he is heralded as the Civil Rights President. How could the lies and deceit run so far and so deep?

President Johnson's words and actions prove that just because someone does something that seems right, doesn't mean they have your best interests at heart. LBJ didn't and today's *Democrat Party* doesn't either! Why are they doing this to you?

The *Democrats* want power, control, and money. When the *Left* looks at you, they don't see humanity; they see dollar signs, they see a ticket to maintain power and control. You are a vote to them; nothing more and nothing less. We see this now extending to the illegal immigrants, and God forbid prisoners, even murders and terrorists who may never see the light of day on the outside of prison walls.

If you dare to disagree with them, they will attack and do their best to destroy you too. Ask Candace Owens or the countless others who have woken up and realized the truth. They had the courage of their convictions, questioned with boldness, and found freedom.

Here's what you have gained in the last two years through Conservative leadership; more jobs, more money in your pocket; greater liberty; lower gas prices; a greater sense of peace and prosperity.

The *current* Administration has accomplished more in two years than all their predecessors. Yet we are led to believe they are racists who want to hold us in bondage. Nothing could be further from the truth.

Like Owens, you have to be willing to look and see the truth to find the freedom already given to you. Like what happened to C. L. Bryant in the late 80s. He stumbled upon a radio talk show and heard a conversation from an entirely different point of view that made him ask questions. Afterward, he became a regular listener of Rush Limbaugh agreeing with his message of lower taxes, less government, and more freedom. Yes, freedom comes when the government is less involved in our daily lives. But the government, at least the liberal, the socialistic government doesn't want you to know this. A government allowed to take the role of benefactor, provider, and burden bearer will eventually become a pseudo-god. What's next to go? Faith or religion. A tyrannical government cannot allow any other source to take its place, including the Creator of the Universe, God almighty.

C. L. Bryant the former democrat radical, NAACP Chapter President, and current Senior Fellow at Freedom Works in Washington DC has been scorned for being bold enough to say that through government hand-outs and social programs, "Government has become 'Daddy.'" And he is right. It might be a harsh truth-pill to swallow but take it anyway. The truth will set you free.

Self-reliance is the key to freedom. Dependence on the government holds people in captivity. You will never have more than the government allows you to have, period.

Like I shared earlier about Margaret Sanger's enlistment of black ministers, the *Left* knows the way into the heart of the masses is through their ministers. If they can capture the mind of the minister or pad his/her hand with a token, then they will win the people who sit under their leadership.

And I need to throw this into the mix too. President Lyndon B. Johnson, who was horrifically racist and spoke openly about his beliefs, also ensnared the Church through his 1954 amendment to silence the church. The Johnson Amendment is a provision in the U.S. tax code, since 1954, that prohibits all 501(c)(3) non-profit organizations and churches from endorsing or opposing political candidates.[16]

While the *Left* doesn't want to tell the truth about who they are and what they have in store for our nation, they also don't want the Church to be involved or have a voice in the political debate. Why in 1954? Because the *Republicans* would have sided with equality and freedom as well as advancement and prosperity for black Americans. Why today? Because the *conservative* point of view sides with God, country, life, Christian values, freedom of religion, free enterprise, and better choices for education.

You see, there have been two constant goals of the *Left*; deceive the black community and silence the truth. But they can do neither because people, no matter their skin color or their socio-economic status, will eventually search for truth. They have enough common sense to look at the facts and see what is going on.

I remember when I woke up! It was 1992. Politically I had already transferred my Party affiliation to Independent because I wanted to think and decide based on my personal convictions. Then the news broke about President Clinton not just harassing women, but raping them. And, he signed into law a "Don't Ask, Don't Tell" policy allowing lesbians, gays, and bisexuals to enlist in the military. It was at that time I realized that the Democratic Platform and Agenda did not align with my Christian values. I switched to the Republican party and never looked back.

They think we're stupid, let's prove them wrong!

Stop playing their game of deception. Decide to be your own man or woman. The only effective weapon they have now is to keep us divided, and they are doing their level best at trying to maintain and even widen the gap. Let's unite and stop them now!

CHAPTER 5

A House Divided

Truth: *A house **united** will not fall.*

As much as there are voices of reason that bring an end to injustice, there are also voices of dissent that further the divide between people.

A house divided cannot stand...

United we will not fall!

The division holds people in a state of confusion, causing them to become unstable in their thinking, their emotions, their beliefs, and their relationships. Instability gives way to chaos. A system thrown into chaos can be reshaped into a structure that controls the everyday life of people. It punishes the dissenter and thereby reinforces in others a fear that silences them, creating further divides.

Just before this book went to press, Joe Biden was throwing his hat in the ring for the 2020 Presidential Election. In his announcement video, the first thing he did was play the race card by replaying a false narrative about the 2017 riots in Charlottesville Virginia. He baited you. Did you take the bait?

Let me clear up this matter right now. President Trump did not say there were good White Nationalists and good neo-Nazis. As usual, the Fake News took his words out of context by not reporting to you the FULL transcript of what the President said. But half the country believes he did and that's the narrative the *Left* keeps pushing. It's a lie. Here's the *unedited* transcript[17]:

> *I think there is blame on both sides. You look at both sides. I think there is blame object both on both sides. I have no doubt about it. You don't have doubt about it either. If you reported it accurately, you would say that the neo-Nazis started this thing. They showed up in Charlottesville. Excuse me. They didn't put themselves down as neo-Nazis. You had some very bad people in that group. You also had some very*

fine people on both sides. You had people in that group — excuse me, excuse me. I saw the same pictures as you did. You had people in that group that were there to protest the taking down, of to them, a very, very important statue and the renaming of a park from Robert E. Lee to another name. [17]

George Washington was a slave owner. Was George Washington a slave owner? So, will George Washington now lose his status? Are we going to take down -- excuse me. Are we going to take down statues to George Washington? How about Thomas Jefferson? What do you think of Thomas Jefferson? You like him. Good. Are we going to take down his statue? He was a major slave owner. Are we going to take down his statue? It is fine. You are changing history and culture. [17]

You had people, **and I'm not talking about the neo-Nazis and the white nationalists. They should be condemned totally.** *You had many people in that group other than neo-Nazis and white nationalists. The press has treated them absolutely unfairly. Now, in the other group also, you had some fine people but you also had troublemakers and you see them come with the black outfits and with the helmets and with the baseball bats. You had a lot of bad people in the other group too.* [17]

Joe Biden has access to the entire press conference transcript. Why would he not tell *you* the truth? Why is Joe Biden continuing to push a false narrative? He wants to inflame you. Then he elevates himself to a saint and champion of Civil Rights.

Biden has said several things that sound racially motivated. Remember what he said about President Obama? *"I mean, you got the first mainstream African-American who is articulate and bright and clean and a nice-looking guy..."* [18] What about Alan Keyes and Herman Cain; both extremely intelligent and articulate? Simple, they are black Republicans and Conservative.

All of this flies in the face of Dr. Martin Luther King Jr., who believed we should judge a man by the content of his character.

It's all about division, and as long as race-baiting keeps us divided, we will never be able to realize our full potential as human beings. And further, the blessing we all want to receive is only released when we dwell together in unity. One nation, under God! This nation isn't perfect, but we can take a stand with our Creator who gave us inalienable rights then lead this nation to freedom in Christ.

America was founded on Judeo-Christian values, based on the blueprint God gave us in the Bible. And it is on those truths that we established our system of law and justice. Legal battles have plagued this nation for decades to forcibly remove monuments and plaques displaying The Ten Commandments from official buildings throughout our nation to erase any references to our founding principles. Again, the blotting out of any reference to God so that the government can take His place.

History has been intentionally revised to erase the truth, to fundamentally change who we are. And, if we don't know where we came from through history, we won't know who we are. At this point, anyone can step in to shape our identity, our thinking, our foundational truths, and our future to suit their evil intent.

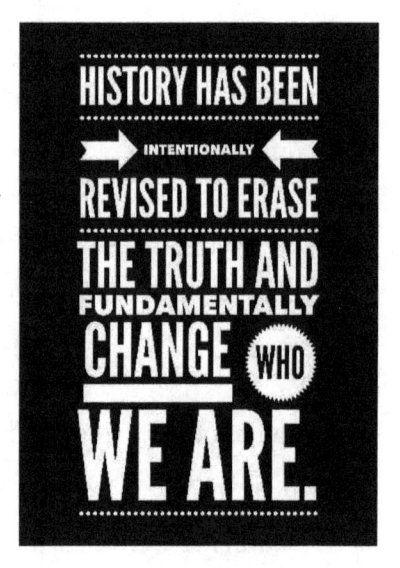

How can we reunify in this great divide? Agreement. There are fundamental issues upon which I believe we can agree. And then we must go back to God's Word because if we don't, we set ourselves *against* Him. And friend, that is not the place we want to be.

Can we agree that life is precious and should be defended? Can we agree that our children should live in safety? Can we agree that we want the best for our children and grandchildren? Can we agree they should have quality education and an opportunity for success? Can we agree that children thrive when both parents are in the home? Can we agree that free enterprise is what makes the American Dream attainable for everyone? Can we agree that working and producing is excellent and needful for all to have a sense of accomplishment and success? Can we agree that a set of boundaries are useful for maintaining peace and order, whether personally or nationally?

There is so much we can agree on that should not be political in nature.

One thing is for sure, a house divided against itself will not stand. But conversely, a house united will never fall. We want to be that house united for the sake of our children because good and blessing come through unity.

We must begin by honoring the truth and the truth-teller.

CHAPTER 6

Honoring the Truth

Truth: *If you don't honor the Truth, you are not honoring Jesus.*

A world renown leader once famously said, "A lie can go half-way around the world before the truth has a chance to put its pants on." It was no doubt true in the 1930s and 40s before technology and the 24-hour news cycle. Can you imagine how fast lies travel today? So fast that the truth has a hard time fighting through the disruptions and outrage created by the lie. In fact, the truth can be so dismissed that it will never be accepted.

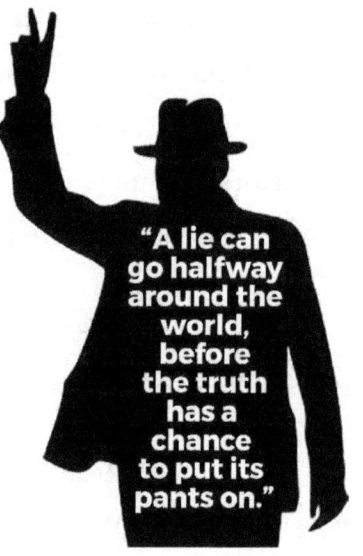

In this high-tech information culture, another anomaly has occurred; people are getting their news, and facts, by soundbite. In 1968 a soundbite was 43 seconds. Today the average length of a soundbite is estimated to be less than 8 seconds. These short bits of information are swaying the minds of people. Unless we take the initiative to ask questions so that we can see beyond the immediate, we end up believing the soundbite is true. Also, leaders who want to influence our thinking have weaponized the soundbite to push their agenda by inserting false information and lies to deceive intentionally. And, making matters worse, we have lost our ability to think critically allowing our emotions to lead our decisions rather than facts. All of this leaves the door wide open to shift a nation's heart from freedom loving to a willingness to give up personal liberties and allow the government full control over their lives.

An example of this is the past two years of the "Trump Collusion with Russia" narrative. A phrase that became a drumbeat on the *Left* and deceived so many people that when Special Counsel Mueller's report was released revealing that there was no collusion by the Trump Campaign, the truth could not be accepted by the *Left*. The facts are still denied in favor of keeping those believing the lie in deception.

Truth must be honored, pursued, and courageously believed, because the end for those who believe the lie is destruction. The fascinating thing about a lie is that it gets bigger and bigger with each push. And as the lie becomes bigger, it is more powerful to deceive and destroy.

To protect ourselves from deception, we must be seekers and lovers of truth. We must love the truth and tell the truth, not according to individual relevancy but to the moral law upon which the foundation of our nation stands. I'm referring to The Ten Commandments here, the Law given by God to Moses for the governing of His people. No wonder many in our nation want any mention of that Law removed from every courthouse and government building. Why? To acknowledge God as the Lawgiver is to put Him above government. The truth is, He is above government. It is God Who ordained government in the beginning.

Those who do not believe there is a God, believe the government is the absolute authority and provider for the people. Their belief is dangerous, a danger that the Founders were well acquainted with through the tyranny of the British King and Britain's rule over the Colonies which ultimately inspired the Declaration of Independence and birth of our nation. The Puritans knew it too. You might remember them as the Pilgrims. They endured religious tyranny and risked their lives to find a place where they could worship God and live free from systematic persecution by both the corrupt church and monarch. Our Founders and our Forefathers both rejected the absolute control of the government and religious persecution. We should reject it too!

An agenda that seeks to control the people and remove all business and private ownership of land and resources cannot have absolute control until the people's reliance on God is removed.

For this reason, most nations ruled by the ideologies of Socialism and Communism are filled with Atheists. As confirmed by rulers like Marx and Lenin, who both believed that faith did little more than prop up the middle class. Karl Marx famously wrote, "Religion is the opium of the people." And Lenin was so afraid of the middle class or "bourgeoisie" he did not want to make the removal of religion the number one focus as he implemented Socialism. Instead, he focused on the benefits of socialism first ensnaring the people. Once they were captured by their thinking that socialism is best for everyone, then God as the ultimate Source and Provider would be removed. At this point, the people would not resist, but joyfully follow the governing ruler straight into bondage. In fact, many would be so deceived they would demand the very things that would make them pauper and prisoner to the one who seeks to destroy.

Hope is so intertwined with faith that a socialist or communist agenda must destroy faith to destroy hope and remove those who believe, to fully capture the people who cannot resist them and their evil plans. How would they resist? Faith in God and prayer would surely defeat their plans because when God's people cry out, He hears, then He answers and delivers them out of trouble.

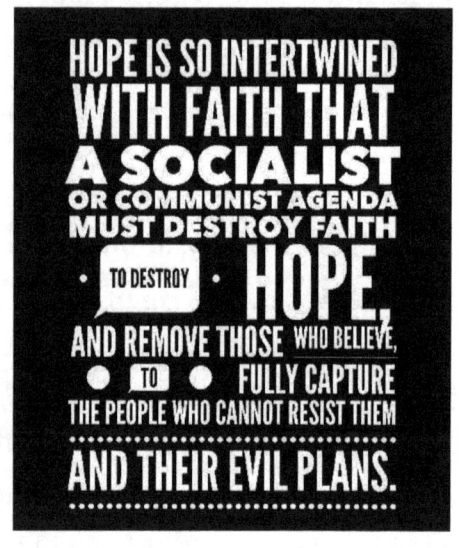

In a culture that lives by soundbites with leaders who understand how to sway public opinion with the least amount of information, it is easy to manipulate the minds of people. It takes determination to search for the truth, and a willingness to hear all sides before an uncompromising decision can be made. It also takes a person willing to live above the fray who can silence the loud voices of dissent to hear from the One Voice that matters.

The deceiver knows when lies are often repeated, they become truth to those who receive them. The lie has been created as an illusion of truth. It seemed right, sounded right, had a ring of truth to it, but in the end, it was still a lie. The problem is that most people don't realize they believed a lie until it is too late. Let that not be you. Let us be courageous for the truth and diligent to guard our hearts because once the lie has taken root, it can become nearly impossible for it to be overcome by the truth. It takes courage to see the underlying reason why the liar desires to deceive in the first place.

Truth is not relevant to a person, an entity, or a nation. There is no such thing as personal truth. Each person has a belief system of his/her choice, but there is only one absolute eternal truth that will stand throughout eternity. And, if we want to remain blessed as a nation, we will honor that truth and accept the One Who gave it.

If you are a Christian, you should honor the truth. If you honor the Truth, Jesus, He will honor you. But if you follow a socially driven pseudo-faith; a compromised, watered down truth that fits an agenda of deception and destruction, then you will easily dishonor the Truth and ultimately lay down your freedom; life, liberty, the pursuit of happiness and Christianity too.

How can you make sure you are on the right side? God's side? Ask yourself a question. Am I supporting things that bring Him glory, honor His name, and further the Truth of His character and promises? Am I living in a way that promotes life, prosperity, and health? Or do I side with the lies that divide with a goal of control?

If you are a Christian, what is your worldview? Do you see everything through the filter of God's character and Word? Or through the filter of division, strife, retribution, retaliation, reparation?

A Christian should have a biblical worldview. If he/she does not, they will follow lies, fables, half-truths, and evil men. They will support issues that are an affront to God and misrepresent the message of love and unity Christ came to demonstrate and release.

Honor the Truth and get on the right side, God's side, of every issue in every matter. And do not give up your freedom in exchange for bondage. The Barna Research Group defines a biblical worldview[19] as;

> #1 **Absolute moral truth exists,** and it only comes from God. He is the first and final authority over the affairs of men.
>
> #2 **The Bible is accurate in all of the principles it teaches.** As believers, we adhere to these principles so that we can live in peace and ensure the tranquility that God promises when we keep His commands.
>
> #3 **Satan is a real being or force, not merely symbolic.** He exists to create strife, divide people, steal, kill, and destroy. He will never be satisfied until he has removed every image and glory of God from people.
>
> #4 **People cannot earn their way into heaven by trying to be good or by doing good works.** It is by grace through faith in Jesus Christ alone.
>
> #5 **Jesus Christ lived a sinless life on earth.** Died on the Cross for our redemption. Those who are Christ's are set free.
>
> #6 **God is the all-knowing, all-powerful creator of the world and still rules the universe today.**

Honoring the truth is a foundational issue. Following corrupt leaders is not uncommon to man. While the words we use and challenges of our time may be different, the problem is the same as those who sailed treacherous waters to find the freedom to worship their God.

A line has been drawn in the sand. Where will you stand? Will you stand with God on His absolute, unchanging truth? I remind myself often; I can't straddle that line for anyone or anything. I can't straddle it for my family, my church, my community, my nation, a political party or a government. And, I'm not afraid of being attacked for believing the truth or being misunderstood. I'd rather know that I'm on the right side, God's side than following man's traditions.

When we dishonor the truth in favor of a lie, we open the door to destruction. As we believe more strongly in the lie, we begin to demand a reckoning for the issue the lie supports and ultimately destroy ourselves. We willingly go along with our captors into a place of bondage. And, the illusion of truth can be much more destructive. We must be able to recognize it.

Our lack of any significant length of attention span has also played into the hands of the deceivers. They know they can shift thinking quickly and easily with a headline. They bank on people not exercising due diligence to seek out the truth, research the facts, and make a choice to go the right way rather than the wrong way.

I believe we are close, maybe within a few short years, to losing our religious freedom. The warnings are everywhere. Have you sensed them? If so, it's time to respond.

CHAPTER 7

The Last Trumpet

Truth: *America is now and always has been a Christian nation.*

We are in a battle—a battle for America.

I believe a trumpet has sounded in our nation, and God is calling His people to come together as one to stand for what is right in His eyes. It may be our last chance. Wouldn't you agree that our God reveals times and seasons through signs and wonders? Of course, He does! He parted the Red

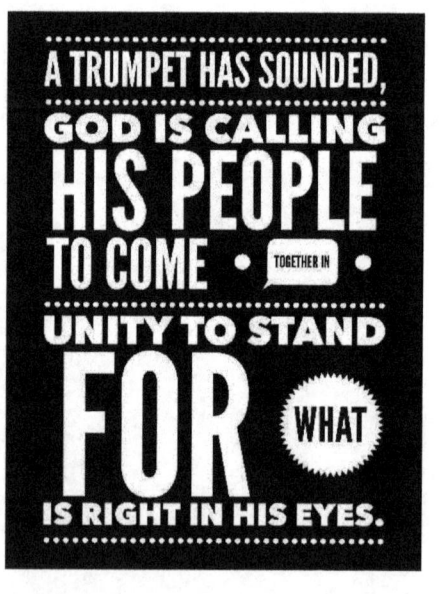

Sea and destroyed His enemies as they sought to attack His people.

In times like these, God will raise leaders who will stand in the gap for His people and His plan. It's like He sticks His finger in turbulent waters to bring peace and prosperity to a time when destruction and oppression seem sure. I believe God did this in 2016. Before I go there let's think about years past.

Never at a time in history was Religious Freedom under attack more than it was during the last *Administration*. President Obama boldly declared America was no longer a Christian nation but a nation of many belief systems and unbelievers. His assertion shocked many, since 235 years of history prove the United States of America is a Christian nation.

Discovering a nation's true identity requires more analysis than merely considering the percentages of people who classify themselves by a particular belief system. Identity is based on a nation's core values and founding principles. America is undoubtedly a Christian nation because our rule of law, ownership of property, equal justice, self-governance, and individual freedoms guaranteed by the Bill of Rights are rooted from Judeo-Christian values.

There are no less than 99 reported examples of hostile acts toward Christianity during that Administration's tenure[20]. Highlights include:

> *February 2009 – The previous Administration announces plans to revoke conscience protection for health workers who refuse to participate in medical activities that go against their beliefs, and fully implements the plan in February 2011.* [20]

> *May 2009 – The White House budget eliminates all funding for abstinence-only education and replaces it with "comprehensive" sexual education, repeatedly proven to increase teen pregnancies and abortions. The deletion is continued in subsequent budgets.* [20]

> *October 2010 – The previous Administration deliberately omits the phrase about "the Creator" when quoting the Declaration of Independence – an omission made on no less than seven occasions.* [20]

> *March 2011 – The previous Administration refuses to investigate videos showing Planned Parenthood helping*

alleged sex traffickers get abortions for victimized underage girls. [20]

April 2011 – For the first time in American history, The Administration urges passage of a non-discrimination law that does not contain hiring protections for religious groups, forcing religious organizations to hire according to federal mandates without regard to the dictates of their own faith, thus eliminating conscience protection in hiring. [20]

February 2012 – The previous Administration forgives student loans in exchange for public service but announces it will no longer forgive student loans if the public service is related to religion. [20]

February 2013 – The previous Administration announces that the rights of religious conscience for individuals will not be protected under the Affordable Care Act. [20]

April 2013 – The United States Agency for Internal Development (USAID), an official foreign policy agency of the U.S. government, begins a program to train homosexual activists in various countries around the world to overturn traditional marriage and anti-sodomy laws, targeting first those countries with strong Catholic influences, including Ecuador, Honduras, and Guatemala. [20]

March 2014 – Maxell Air Force Base suddenly bans Gideons from handing out Bibles to willing recruits, a practice that had been occurring for years previously. [20]

March 2015 – A highly decorated Navy SEAL chaplain was relieved of duty for providing counseling that contained religious views on things such as faith, marriage, and sexuality. [20]

October 2016 – The previous Administration threatens to veto a defense bill over religious protections contained in it. [20]

Many of these actions emboldened the LGBTQ Movement. Activists challenged hardworking Americans whose convictions would not allow them to be involved in activities and events that celebrated and promoted sin. Same-sex couples targeted Christian owned bakeries who refused to bake their wedding cake. In 2015 an Oregon bakery was fined $135,000 by the Oregon Bureau of Labor for refusing to bake a cake for two women. In Colorado, another baker declined based on religious beliefs, and the couple took their case to the US Supreme Court only to lose in 2018.[21]

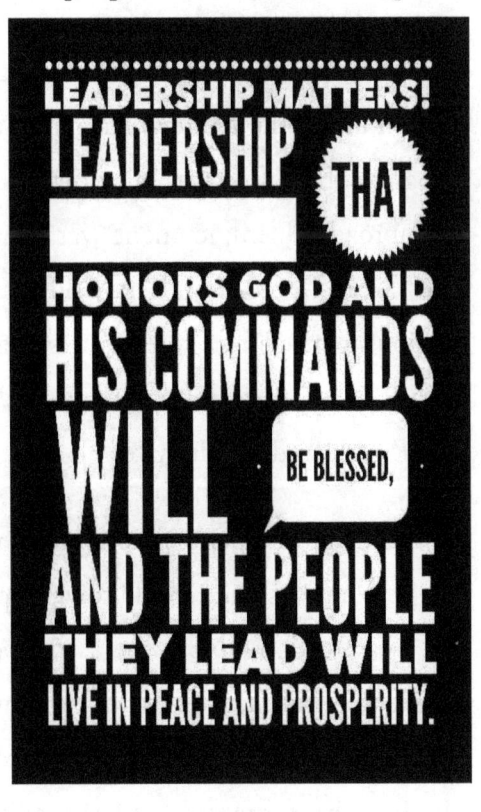

In 2014 Houston Mayor Anise Parker, a lesbian, tried to intimidate five pastors who opposed her "HERO" Bathroom Bill allowing transgender people to use whichever public bathroom they chose.[22] The Bill made it legal for a transgender man to enter a woman's bathroom on public property. When the pastors opposed the bill and secured the petitions required to bring a vote before the people of Houston, the Mayor subpoenaed all their "speeches, presentations, or sermons related to [the equal rights ordinance], the petition, Mayor Annise Parker, homosexuality, or gender identity." Her actions and the bill created a national firestorm, and the Mayor eventually withdrew her subpoena. It was a clear violation of free speech and religious freedom.

Leadership matters! Leadership that honors God and His commands will be blessed, and the people they lead will live in peace and prosperity.

When we look back at the religious persecution and turmoil under the *previous* Administration, it is not hard to recognize the nation was in a downward spiral both at home and abroad. Unemployment was at an all-time high since The Great Depression. The Middle East was on fire, and religious persecution rose at a staggering rate. Christian historical sites were destroyed, and believers were massacred as ISIS declared a Caliphate.

From the moment national leadership shifted in 2017, the pendulum swung back toward peace, prosperity, *and* religious freedom. Now the *Left* doesn't want anyone to believe this, but facts don't lie. Employment is at a fifty year historical high, industry is returning to America, millions of jobs have been created, and the religious liberty that was under fire by the *previous* Administration has been restored. A few highlights here;

April 2017 – The current Administration signed into law H.J. Res. 43, overturning a midnight regulation by the previous Administration, which prohibited States from defunding certain abortion facilities in their Federally funded family planning programs.[23]

May 2017 – The current Administration signed an executive order to greatly enhance religious freedom and freedom of speech, taking action to ensure that religious institutions may freely exercise their First Amendment right to support and advocate for candidates and causes in line with their values; and ensuring that religious Americans and their organizations, such as the Little Sisters of the Poor, would not be forced to choose between violating their religious beliefs by complying with the ACA's contraceptive mandate or shutting their doors.[23]

October 2017 – Under the leadership of the current Administration the Department of Justice issued twenty principles of religious liberty to guide the Administration's litigation strategy to protect religious freedom.[23]

January 2018 – The current Administration <u>reversed</u> the previous Administration's policy denying disaster aid to *houses of worship, allowing houses of worship to receive crucial aid in times of crisis.* [23]

January 2018 – Under the leadership of the current Administration, the Department of Health and Human Services (HHS) announced three major policy changes to protect freedom of religion: forming a new Conscience and Religious Freedom Division, providing HHS with the focus it needs to more vigorously and effectively enforce existing laws protecting the rights of conscience and religious freedom; and proposing to more vigorously enforce 25 existing statutory conscience protections for Americans involved in HHS programs, protecting Americans who have religious or moral convictions related to certain health care services. [23]

Leadership DOES matter! But leaders cannot wholly fulfill what God wants to do in the heart of people throughout this nation. Ultimately, we need revival; we need to return to God.

The church got a reprieve from religious persecution in 2017. I believe this season of heightened religious freedom will be extended to 2024. But what will we do with this time we've been given? It's a short season of time to rise, unify, and do what God created the church to do. And, let's face it,

I'm not talking about a building here. I'm talking about all the amazing people who call themselves Christians. It's time for us to do something!

What needs to be done? First, unity and love need to be restored. Hopeful expectation about a future God has promised needs to be reborn in the heart. His plans for us and our nation are good. He wants to prosper America, which includes you and me. Trust for and loyalty towards our neighbor and community. And together, we need to stand firm on God's Word. Regarding the deceived, we need to have a sense of urgency and determined diligence toward breaking their bonds of deception and setting them free. How? By courageously telling them the truth.

Patrick Henry famously declared, "Give me liberty or give me death." I must admit at the end of 2018, I was crying out to God for the same thing. I realize now that we must all come to a place where we'd rather die than continue on the path, we have laid for ourselves to now. It must be that we will do anything we can to change the course of our nation. Can we turn it around? Yes! With God's help. We must trust Him, align ourselves with His plans, then reject everything culturally and politically that is contrary to His truth.

I'd rather die than be shackled to poverty and perversion. I rather die than go into bondage and godlessness by Socialist rule. I'd rather die than lose the freedom and liberty Christ came to give, and our Forefathers fought and died for to give birth to a nation established on Judeo-Christian values.

I want to live the dream Dr. Martin Luther King Jr. envisioned for all people in America. I believe you do too. His "dream" *is* obtainable right now. We only need to decide what that dream looks like for us, pray, believe, and work hard to achieve it. For now, it's time to let go of the lies.

> I BELIEVE YOU ARE HUNGRY FOR THE VERY THINGS THAT MAKE US GREAT—OUR LOVE FOR GOD, EACH OTHER, AND OUR COUNTRY.

I believe you want to know the truth. I believe you want to be free from division, violence, and chaos. I believe you are hungering for the very things that make us great — our love for God, each other, and country.

Let's take a stand together. Let's herald the truth. Let's be ready when that ultimate last trumpet sounds. Let's join hands now and go into the dark places to bring light. Let's get the treasure God has hidden for us there in the darkness. Let's turn the light on for them so that they can find their way out to freedom too.

May the words of our President ring true and resonate deep in our heart as we come together and take back our nation for the glory of God.

"We know that faith and family, not government and bureaucracy, are the center of American life. And above all else, we know this: In America, we don't worship government; we worship God." — President Donald J. Trump

I believe President Trump is a prophetic sign of the last trumpet. God is always giving us signs about the future. Since taking office, the President has been exposing and tearing down corruption, like every great leader before him. He is God's representative.

CHAPTER 8

Dedicated & Determined

Truth: *I will fight for YOU!*

I was honored in 2018 to receive the Statesmen of the Year Award from the Tennessee State Republican Party and am currently an elected official who serves on the State Executive Committee for TN33. For many years, I have offered a staunch voice for the values that make America great, life, liberty, and the pursuit of happiness. I am passionate about speaking truth and inspiring people to live in the freedom that our ancestors fought for and died to protect. Freedom to pursue prosperity;

to create, build, and establish businesses that benefit our fellow citizens; and to worship our God freely.

I was born and raised in Memphis, Tennessee, obtained most of my schooling through the Memphis public school system before busing was introduced. I'm so proud of my father, now deceased, who was a Baptist preacher. I understood what it meant to be the daughter of a minister and was never offended when called a 'PK' (Preacher's Kid). One of the things I'm most thankful for is that my father required his children to memorize Scripture. It was the best gift I ever received from him. Even to this day, it is God's Word that steers my life and decisions.

I was the only one in my family to graduate from college, graduating from Christian Brothers University in Memphis, where the "Brothers" were able to begin classes with much-needed prayers. I worked an average of 14 hours day and night while attending classes during the last few years of my college career. Those days provided many character-building opportunities and taught me the real value of education *and* hard work.

While I grew up in extreme poverty, I did not realize my family was poor. At one point, we lived in a one-bedroom apartment, my parents, and seven siblings. Unfortunately, my parents divorced, and my mom remarried increasing the family to twelve. Thankfully, we moved into a two-bedroom apartment at that time. As you can see, I understand the challenges people face today; both poverty and the hardships of broken families have touched my heart deeply.

I've known success and prosperity. I enjoyed a long career as a Technology Project Manager until 2000, when the company I worked for outsourced my job to an overseas company. A trend among major corporations in the 90s and 2000s, so it made finding another job in my field extremely difficult. To make matters worse, I fell victim to the Housing Market Crash in 2008 and lost my home. Poverty struck me a second time and then homelessness, but I was determined not to give in to my circumstance. I found success *again*. After networking and pulling myself up, while refusing government subsidies, I am now an entrepreneur who owns a small business providing home healthcare to seniors and jobs to make my community a better place. I'm grateful every time I sign the front of my employees' paychecks, knowing I am providing them with an opportunity to work.

During those challenging years, my world was rocked, but a new awareness was borne in my heart as I began to see how government leaders' decisions were adversely affecting hard working American citizens in such detrimental ways. As a result, I decided to do something! In 2010, I threw my hat in the ring and ran against the entrenched Democratic Representative Steve Cohen. I have been dedicated since that moment to do everything I can to enlighten the hearts and minds of the people in my district as to the dangers of destructive policies, the truth about Socialism and the *Left*'s agenda in hopes of championing the freedom of the people I love so dearly.

I am proud to be a steadfast supporter of President Donald J. Trump. I stood in the rain holding the "Trump/Pence" signs and organized rallies during the Tennessee elections because I believed Trump would win to "Make America Great Again."

> *"Nothing worth doing ever, ever, ever came easy. Following your convictions means you must be willing to face criticism from those who lack the same courage to do what is right. And they know what is right, but they don't have the courage or the guts or the stamina to take it and to do it."* ~ *Donald J. Trump*

I firmly believe that when we vote *our shared values*—a belief in limited government, personal responsibility, and life—we can't lose. It is our common ground on which we can proudly stand.

Now—it's up to YOU!

Do your part to influence those you lead, or *do life* with, in your home, community, or church. Tell them the truth!

In closing, I want to share a prayer for you in hopes you will join me in my quest to herald God's truth and call our family, neighbors, communities, and churches back to Him.

May God grant you the courage to stand, the boldness to speak, and the determination to finish strong in His truth as you honor His name. May you be the answer and not the argument, the one with heavenly wisdom rather than baseless rhetoric, the influential leader rather than a blind follower of lies. May you rise up in this hour to the greatness you were created for and lead people out of bondage physically, emotionally, and spiritually as you too become a "herald of truth." In Jesus name, I pray. Amen.

End Notes

1. "The Ignored story of 'America's Biggest Serial Killer'" | | *Washington Post. George Will.* Web. 7 December 2018 https://www.washingtonpost.com/opinions/the-ignored-story-of-americas-biggest-serial-killer/2018/12/07/6379964a-f986-11e8-8c9a-860ce2a8148f_story.html?utm_term=.de985e0acae8
2. "Rules for Radicals" | | *The Citizens Handbook. Charles Dobson.* Web. http://www.citizenshandbook.org/rules.html
3. "' Empire' star Jussie Smollett: Attackers Yell 'This is MAGA Country' during beating" | | *USA Today. Sara M Moniuszko and Jayme Deerwester.* Web. 29 January 2019 https://www.usatoday.com/story/life/people/2019/01/29/empire-star-jussie-smollett-assaulted-possible-homophobic-attack/2709986002/
4. "Judge scolds Jussie Smollett over allegations he staged racist, anti-gay attack: 'Vile and despicable'" | | *Chicago Tribune. Megan Crepeau, Jason Meisner, and Jeremy Gorner.* Web. 21 February 2019 https://www.chicagotribune.com/news/local/breaking/ct-met-jussie-smollet-arrest-20190221-story.html
5. "The Negro Project: Margaret Sanger's Eugenic Plan For ..." | | *Concerned Women for America.* Web. 10 May. 2019 https://concernedwomen.org/the-negro-project-margaret-sangers-eugenic-plan-for-black-americans/.

6. "The Negro Project: Margaret Sanger's Eugenic Plan For ..." | | *Concerned Women for America*. Web. 10 May. 2019 https://concernedwomen.org/the-negro-project-margaret-sangers-eugenic-plan-for-black-americans/.
7. "Margaret Sanger, Race and Eugenics: A Complicated History ..." | | *Time. Jennifer Latson*. Web. 14 October. 2016 http://time.com/4081760/margaret-sanger-history-eugenics/
8. "Ohio Democrat Pushes for Black Children to Be Aborted." | | *The Daily Caller. Grace Carr*. Web. 10 April. 2019 https://dailycaller.com/2019/04/10/ohio-heartbeat-bill-exemption/
9. "Planned Parenthood Targets Minority Neighborhoods" | | *Protecting Black Life. Executive Summary*. Web. 10 May. 2019 https://www.protectingblacklife.org/pp_targets/index.html
10. "The Abortionist's Eye Is on Us" | | *The Washington Times*. Web. 20 July. 2009 https://www.washingtontimes.com/news/2009/jul/20/the-abortionists-eye-is-on-us/.
11. "Abortion and Race" | | *Abort73. Loxafamosity*. Web. https://abort73.com/abortion/abortion_and_race/
12. "Abortion – Black Genocide? 'My People Are Dying' | | *BreakPoint. Chuck Colson*. 14 January. 2011 http://www.breakpoint.org/2011/01/breakpoint-abortion-black-genocide/

13. "Intolerant, Godless Democrats Boo Inclusion of God, Israel In Party Platform" | | *C-Span*. Web. 5 September. 2012 https://www.c-span.org/video/?c3872932/intolerant-godless-democrats-boo-inclusion-god-israel-party-platform
14. "The Inconvenient Truth About the Democratic Party" | | *Prager University*. Dennis Prager. Web. Video. 22 May. 2017 https://www.prageru.com/video/the-inconvenient-truth-about-the-democratic-party/
15. "Leftism Debunked: The 'Southern Strategy'" | | *Z. Chase Watkins*. Web. 7 December. 2018 http://www.aviewfromgenz.com/leftism-debunked-the-southern-strategy/
16. "Johnson Amendment" | | *Wikipedia*. Web. 10 May. 2019 https://en.wikipedia.org/wiki/Johnson_Amendment.
17. "Full transcript: Donald Trump's press conference defending the Charlottesville rally" | | *Vox*. Libby Nelson and Kelly Swanson. Web. 15 August. 2017 https://www.vox.com/2017/8/15/16154028/trump-press-conference-transcript-charlottesville
18. "Joe Biden's Description of Obama draw Scrutiny" | | *CNN.com*. Web. 9 February. 2007 http://www.cnn.com/2007/POLITICS/01/31/biden.obama/.
19. "Barna: How Many Have a Biblical Worldview?" | | *Christianity Today*. Ed Stetzer. Web. 9 March. 2009

https://www.christianitytoday.com/edstetzer/2009/march/barna-how-many-have-biblical-worldview.html

20. "America's Most Biblically-Hostile President" || *WallBuilders.com. David Barton.* Web. https://wallbuilders.com/americas-biblically-hostile-u-s-president/

21. "Supreme Court rules in favor of baker in same-sex wedding cake case" || *CBS News.* Web. 4 June. 2018 https://www.cbsnews.com/news/supreme-court-rules-in-favor-of-baker-who-denied-same-sex-couple-a-wedding-cake/

22. "City of Houston Anti-Discrimination HERO Veto Referendum, Proposition 1" || *BallotPedia.com.* Web. November 2015. https://ballotpedia.org/City_of_Houston_Anti-Discrimination_HERO_Veto_Referendum,_Proposition_1_(November_2015)

23. "President Trump Has Been A Champion for Religious Freedom ..." || *WhiteHouse.gov.* Web. 10 May. 2019 https://www.whitehouse.gov/briefings-statements/president-trump-champion-religious-freedom/.

Printed in the USA
CPSIA information can be obtained
at www.ICGtesting.com
CBHW050115310824
13726CB00117B/1338